AF228876

SPACE EXPLORATION

# COLONIZING MARS

by Tammy Gagne

BrightPoint Press

San Diego, CA

© 2023 BrightPoint Press
an imprint of ReferencePoint Press, Inc.
Printed in the United States

For more information, contact:
BrightPoint Press
PO Box 27779
San Diego, CA 92198
www.BrightPointPress.com



LIBRARY OF CONGRESS CATALOGING-IN-PUBLICATION DATA

Names: Gagne, Tammy, author.
Title: Colonizing Mars / by Tammy Gagne.
Description: San Diego, CA : BrightPoint Press, [2023] | Series: Space Exploration| Includes
    bibliographical references and index. | Audience: Grades 10-12
Identifiers: ISBN 9781678204266 (hardcover) | ISBN 9781678204273 (eBook)
The complete Library of Congress record is available at www.loc.gov.

# CONTENTS

- On average, Mars is located about 140 million miles (225 million km) from Earth. It takes spacecraft about seven months to make the journey.

- Landing a crewed spacecraft on Mars will be extremely challenging.

- Supplies for a Mars colony will first need to be delivered from Earth. But eventually Mars colonists will find and use resources on the planet.

- Colonists on Mars will need energy sources on the planet. They might use water, wind, and sunlight to create electricity.

- The environment of Mars poses many dangers for colonists. They will face high levels of radiation. Dust storms may damage equipment and block sunlight from reaching solar panels.

- Scientists have several ideas about how to design a colony on Mars. Large glass structures could protect colonists from the Mars environment. Underground structures are another possibility.

- Colonists on Mars will need to grow their own food supply. They may grow corn and soybeans in greenhouses. They may raise insects to be eaten as a source of protein.

# INTRODUCTION

# GREENHOUSES ON MARS

"I love watching people lift things on this planet," says Tom. He and his coworker Maria are eating lunch on Mars. A nearby construction crew builds a new greenhouse for the colony. An engineer picks up an entire section of the greenhouse all by herself. The gravity on

Mars is about one-third as strong as the gravity on Earth. Weaker gravity makes lifting heavy objects easier.

"It does take some getting used to," Maria replies as she eats her salad. "These

tomatoes are even tastier than the last

batch." Tom and Maria are plant scientists.

They grow food for the Mars colonists. The

tomatoes were grown in a greenhouse.

Greenhouses are necessary on the

planet. Mars is too cold and dry for crops

to be grown outdoors. The colonists spend

little time outside the compound. When

they do go outdoors, they have to wear

protective suits. They use oxygen tanks

to breathe.

"Did you notice that these tomatoes

are bigger, too?" Tom asks. "I think I finally

figured out the ideal soil mixture."

*Mars is nicknamed the Red Planet because of its red color.*

Tom and Maria miss some things about Earth. But they have dreamt of this opportunity for years. They are two of many scientists in the colony. They continue to learn more about Mars.

*The Mars rover Perseverance landed on Mars in February 2021.*

## GREAT EXPLORATIONS

People do not live on Mars yet. But by

2022, the National Aeronautics and Space

Administration (NASA) had sent five rovers

to Mars. These robotic vehicles explore the

planet's surface. They send photos and other data back to Earth.

NASA hopes to send the first human mission to Mars in the 2030s. Humans will learn a lot about the Red Planet. The plan is to build a colony on the planet. This will make it possible for people to live on Mars. Much work is still needed for humans to travel to the planet. But a colony on Mars would expand the possibilities of space exploration.

# 1

# HOW WILL HUMANS GET TO MARS?

Traveling in space is challenging. The distance between Earth and Mars varies as the planets orbit the Sun. On average, Mars is about 140 million miles (225 million km) from Earth. It will take spacecraft about seven months to make the journey.

Large amounts of fuel are needed to send supplies to Mars. Rockets will have heavy **payloads**. They need to carry many supplies for people to build a colony on Mars. Colonists will need construction materials. They will also need oxygen, food, and water.

*It takes a lot of fuel to launch a rocket from Earth to Mars.*

Rocket engines work by burning propellants. They release a lot of energy in a short time. However, propellants are heavy. The ion engine is another type of engine. These engines use electricity and a lightweight gas. There is no need for large propellant tanks. Ion engines can fire for a much longer time than rocket engines.

But they push with less force. It could take two years for a rocket using an ion engine to reach Mars. This type of engine could be great for cargo spacecraft. The cargo ship could carry supplies to Mars ahead of the astronauts.

## LANDING ON MARS

Landing safely on Mars is another challenge. Spacecraft need to slow down to land safely. Earth has a thick **atmosphere**. It is made of gases such as oxygen and nitrogen. They push against a spacecraft as it enters the planet. This helps slows the

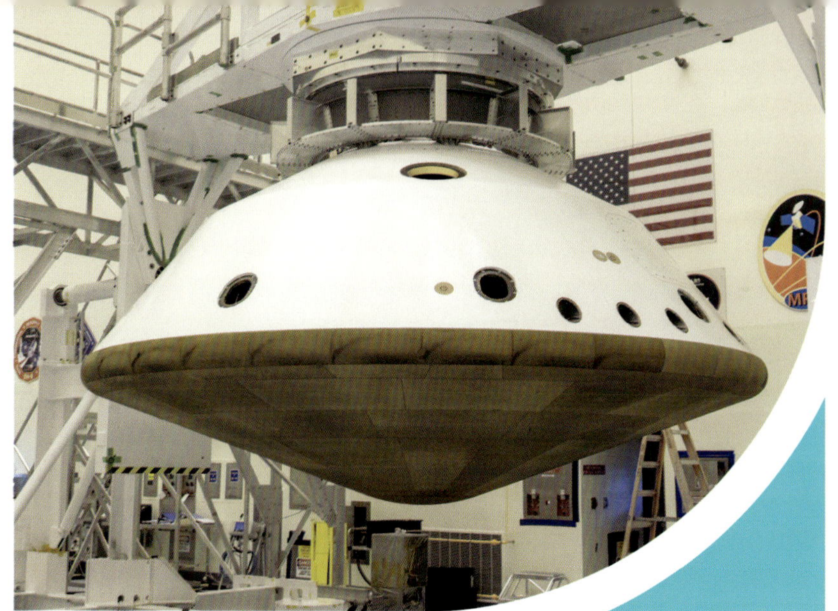

*Spacecraft that land on Mars need a heat shield.*

spacecraft down. It also creates a lot of

heat. The spacecraft needs a heat shield

to survive.

The Martian atmosphere is much thinner.

It cannot slow the spacecraft as much

as on Earth. There is not enough air. But

spacecraft still need heat shields when

entering the atmosphere. Slowing down

safely is a major challenge.

*The Curiosity rover tested a new landing system. It first slowed down with a parachute. Then a sky crane lowered the rover to the ground.*

Scientists have learned a lot about landing spacecraft on Mars. NASA successfully landed its *Curiosity* rover on Mars in 2012. A parachute helped slow the descent. Then, a sky crane with rocket engines gently lowered the rover to the ground. In 2016, Europe's *Schiaparelli* **probe** also made it to Mars. But its parachute did not slow it down enough. The probe was destroyed when it crash-landed. It was moving at 186 miles per hour (300 kmh) when it hit the surface.

Landing spacecraft on the planet remotely is difficult. The task will not be easy for astronauts either. But a crewed mission could control its own entry. It would also receive help from Earth as needed.

## NASA'S TRACK RECORD

NASA has had more successful missions to Mars than any other space organization. As of 2021, ten robots have landed successfully on the planet. Seven of them belonged to NASA. China landed its first rover on the planet in 2021.

Dr. Firouz Naderi has worked on two rover missions to Mars. Naderi said, "One colleague describes the entry, descent, and landing as 'six minutes of terror.'"[1] This is because signals from Mars take a long time to reach Earth. Scientists must wait several minutes to learn whether the spacecraft landed safely.

## THE COST OF THE MISSION

The trip to Mars could cost up to $500 billion. Some people think this price is too high. They argue that rovers offer plenty of information about Mars. Rovers have captured many images on the Red Planet. They also record information about the soil and climate. Retired NASA astronaut Steve Swanson said, "I agree the rovers on Mars have done wonders and helped get people excited about planetary exploration, but I'm sure the excitement of humans going to Mars would be much greater."[2]

Still, the cost must be covered somehow. Colonizing the planet will be even more expensive. SpaceX is one company that plans to build a colony on Mars. Elon Musk is the founder of SpaceX. He thinks the travelers themselves could help fund the

*In the future, it may be possible for people who are not professional astronauts to live on Mars..*

project. He has suggested making loans to the colonists for the trip. They would then work in the colony to pay off the loans. Musk thinks the cost of living on Mars will be more affordable over time. His goal is that people may be able to move to Mars for less than $100,000.

# 2

# GETTING RESOURCES ON MARS

The first humans on Mars will bring supplies to set up the colony. They will still depend on supplies from Earth after the colony is built. Resources such as oxygen and water are limited on Mars. But sending additional supplies from Earth will take a long time. It will also be expensive. Scientists hope that the Mars

colony will one day be self-sufficient. This

means the colony will no longer rely on

supply missions. Instead, colonists will

*Astronauts in space depend on resupply missions. A resupply vehicle carried supplies to astronauts in the International Space Station in 2003.*

use resources already on Mars to fulfill their needs.

A large amount of frozen salt water lies near the planet's poles. More water is underground. Research suggests that Mars's crust could contain as much salt water as half of the Atlantic Ocean. Scientists could use this resource to make fresh water for colonists. Many people believe that a Mars colony is possible because of this water supply. Colonists could use water that is already on the planet. Resupply missions to Mars would not need to deliver water to the colonists.

# ICE ON MARS

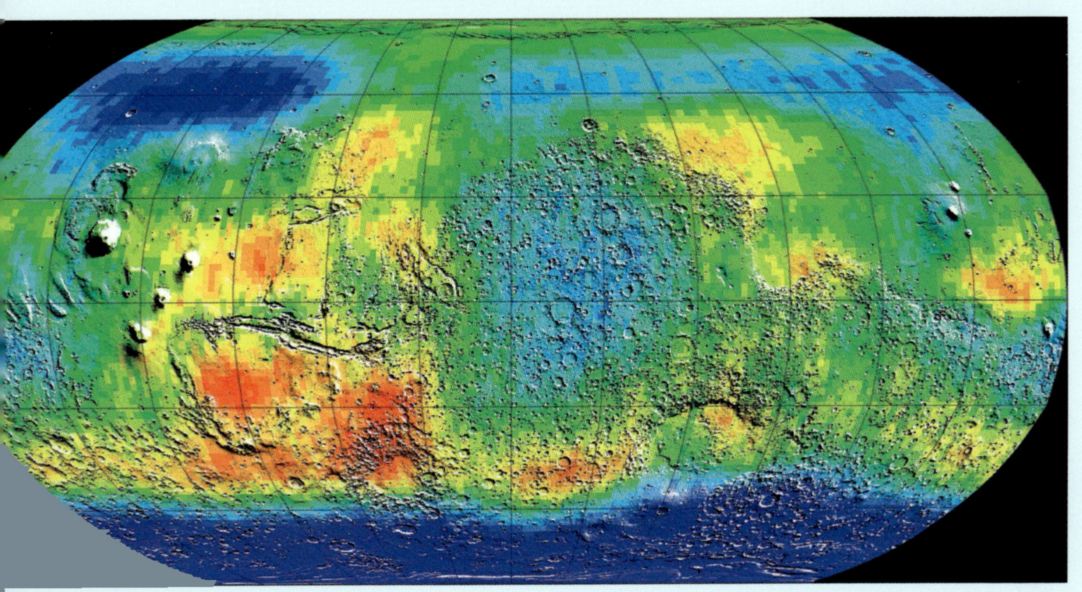

*Spacecraft on Mars have searched for signs of hydrogen, an element in water. Areas with this element may have ice. Blue shows high amounts of hydrogen; green, yellow, and red areas have less.*

These missions could occur less often.

There would also be more room on

cargo ships for other materials. Scientist

Christopher Adcock researches Mars.

He says, "The water we need for long-term human missions to Mars might be right at our feet when we get there."[3]

This water could also be used to restore oxygen supplies. Water is made of hydrogen and oxygen. Scientists could break apart water **molecules**. This would make breathable oxygen.

## GROWING THEIR OWN FOOD

For the colony to be self-sufficient, colonists will need to grow their own food. The climate on Mars is cold and dry. Crops will need to be grown in greenhouses.

*Experiments on the International Space Station help scientists learn about growing crops in space.*

Greenhouses protect the plants from these conditions. They keep air warm and moist. Greenhouses are made of glass. Sunlight can shine through this material. But Mars receives less energy from sunlight than Earth does. Colonists may also use special lighting that mimics sunlight.

Scientists have experimented on Earth. They wanted to know if it would be possible to grow crops in Martian soil. Scientists **simulated** Martian soil in a lab. They received promising results. They were able to grow a variety of vegetables in the soil. They successfully grew peas, radishes, and tomatoes. These crops have seeds that can be saved and planted later. This will allow colonists to keep growing those foods.

It is an important part of developing a

self-sustaining colony.

## FINDING ENERGY SOURCES ON MARS

Colonists will need energy for light and heat.

They also need to power machinery. Much

of the electricity on Earth is produced by

coal or oil. These resources are unavailable

on Mars. The colonists will need to rely

on renewable energy sources. Scientists

study how to make these energy sources

more effective.

*The Sojourner rover was powered by solar energy.*

The colonists will get some energy from solar panels. Some rovers on Mars are powered by solar energy. But these machines do not require much energy. A colony needs much more power. Solar energy alone will not be enough. Sunlight is only 43 percent as strong on Mars as it is on Earth. There are also frequent dust storms on Mars. They block sunlight. Solar panels cannot create power without sunlight. Solar panels also do not work at night. Batteries could help colonists store electricity and use it later. Solar energy will answer some of their power needs. But they will need other energy resources as well.

Wind energy is another option. People use wind turbines to produce energy on Earth. These structures could be used during dust storms on Mars. They would continue to produce energy at night.

However, wind energy is less effective on Mars than on Earth. This is because the atmosphere is very thin. It is about 1 percent as dense as Earth's atmosphere. This means even a fast wind carries very little force.

Some experts think that water is the best energy resource on Mars. Scientists can break down water to produce hydrogen. This gas could power machinery in the colony. Water can also be used to produce methane. Methane is a part of some rocket fuels. Colonists could create fuel on Mars for return trips to Earth.

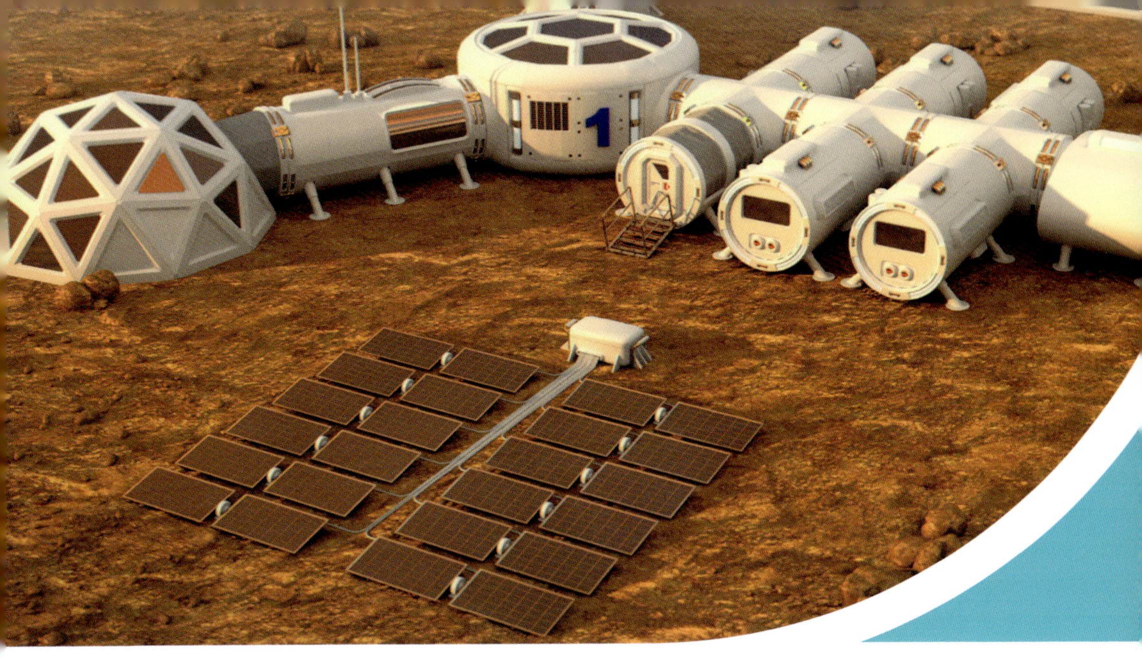



*Solar energy will be one energy source on Mars.*

All energy resources will need to be managed carefully. Resources on Mars are limited. Colonists will also need to monitor all of their supplies. They need to make sure they do not run out of food and water. They must keep a close eye on the amount of oxygen they have. These supplies are necessary to maintain a colony on Mars.

# 3

# THE DANGERS OF MARS

Mars is a dangerous planet. The environment is much different from that of Earth. Living on the planet will expose colonists to many new dangers. Radiation is one of the biggest threats. The Sun emits powerful radiation. Long-term radiation exposure can lead to serious health problems and even death.

Earth has a thick atmosphere. It also has a strong magnetic field. These features help protect humans from the Sun's radiation.

But Mars has neither. Radiation levels on Mars are extremely high as a result.

Colonists will need to wear space suits whenever they go outside. This will shield them from some radiation. But other safety measures also need to be taken. For example, colonists may need to limit the time they spend outdoors. This will help keep their radiation exposure as low as possible.

One long-term way to reduce radiation is to change the Martian atmosphere. Scientists call this terraforming. Large amounts of carbon dioxide are frozen at

*Scientists hope that indoor spaces on Mars will have a similar environment to Earth.*

the poles of the planet. Melting the carbon

dioxide would release it as a gas. The gas

would create a thicker atmosphere. It would

shield the planet from radiation. Colonists

on Mars could grow plants that would take

in carbon dioxide and release oxygen.

People could breathe on Mars if the planet

had a thick atmosphere with oxygen. But a huge amount of plants would be needed. Terraforming would take decades to accomplish. Scientists are not even sure it is possible. If successful, Mars colonists could go outside without protective clothing.

# COLD TEMPERATURES

The temperatures on Mars are another risk to colonists. The average temperature on the planet is –80 degrees Fahrenheit (–62°C). Temperatures can vary greatly throughout the day. Summer temperatures near the planet's equator can reach 70 degrees Fahrenheit (21°C) during the day. The area may get as cold as –100 degrees Fahrenheit (–73°C) at night. NASA scientist Pascal Lee said, "[Freezing temperature] is something that would kill you over the course of hours if you are not properly warmed."[4]

*Ice on Mars is buried beneath the surface.*

Temperatures are more extreme close to the poles. Winter temperatures there can dip as low as –225 degrees Fahrenheit (–143°C). Avoiding this region would protect colonists from the cold. But colonists could also benefit from the ice located there. The ice could be an important water supply.

## DUST STORMS

Dust storms on Mars can become so large that they surround the entire planet. Some storms can even be seen from Earth with a telescope. Winds from Martian dust storms can reach 60 miles per hour (97 kmh). The thin atmosphere means these winds are unlikely to cause serious damage to equipment. However, dust storms could make it hard for astronauts to see. They would need to stay indoors during a storm.

*Dust storms on Mars can transform the entire planet. These photos compare the same area of Mars before a dust storm (left) and after a dust storm (right).*

Dust storms will affect colonists in other ways. The dust traps radiation that would otherwise escape into space. Radiation levels on the planet would increase. This can cause health issues.

Scientists are still learning about the planet's dust storms. Jim Watzin heads the

Mars Exploration Program at NASA. He said, "We really need to understand these storms to the degree that we can have some level of forecasting ability."[5] Scientists are not yet able to predict when or where these storms will occur. This makes it hard to prepare for them.

## MADE FOR THE MOVIES

*The Martian* is a popular science fiction movie about an astronaut on Mars. The main character encounters numerous setbacks from dust storms on the planet. The film was a big hit. But NASA scientists say it is unlikely that a dust storm could do the kind of damage viewers see in the movie.

# 4
# EVERYDAY LIFE ON MARS

Scientists are studying Mars to determine the best location for a colony. It should be a place where it is easy to land. Colonists should be able to easily access natural resources. Scientists have discovered several options.

NASA has used rovers to explore Mars. Most of the rovers have landed on the northern **hemisphere** of the planet. There is a lot of flat land on this half of

*It will be easier to build a Mars colony on a flat area, such as in a crater.*

the planet. The southern hemisphere has
more mountains.

The Arcadia Planitia is a possible option
for a colony. These lowlands are in the
northern hemisphere. Research suggests
that this area received large amounts of
snow tens of millions of years ago. Dust
has buried the snow over time. But Mars
colonists could drill underground. Then they
would have access to water.

Deuteronilus Mensae is another option.
Mars rovers have discovered underground
ice there. The ice may be tens of millions
of years old. Colonists could mine for

water. Deuteronilus Mensae is also located

in the northern hemisphere. It is east of

Arcadia Planitia.

# WHAT WILL THE FIRST MARS COLONY LOOK LIKE?

Shelters on Mars will need heavy

**insulation** to trap heat and keep out the

cold. They will also need radiation shielding.

Some scientists think that most living space

should be built underground. The soil would

trap heat. It would also block radiation.

Others think that large domes with multiple

layers of glass would work best. The

material would block harmful radiation.

But sunlight would still pass through.

The sunlight would help colonists grow

plants indoors.

*Glass structures will allow sunlight to shine through. This helps make it possible to grow plants indoors.*

Structures will need to be airtight.

They need to be kept as close to Earth's

environment as possible. Colonists will

spend much of their time indoors. Scientists

hope to build indoor spaces that look and feel like outdoor spaces on Earth. In 2019, Elon Musk said, "For having an outdoorsy, fun atmosphere, you'd probably want to have some faceted glass dome, with a park, so you can walk around without a suit."[6]

## AIR LOCKS

All shelters on Mars will need air locks. Air locks keep oxygen from escaping the shelters when people go outdoors. Colonists will step through a door into the air lock. They wait in this area until that door seals shut. They put on a space suit with its own air supply. Then they can open a second door that leads outside.

# DEALING WITH THE DUST

Particles left behind from dust storms on Mars will be an ongoing problem for colonists on the planet. Dust can damage equipment. NASA has been dealing with this issue for many years with its rovers on the planet. Scientist Michael Smith said, "The dust coats everything, and it's gritty; it gets into mechanical things that move, like gears."[7]

Dust storms will also create energy challenges. Shelters on Mars will have solar panels. Storms will block sunlight. This limits the amount of energy the panels can produce. Dust will continue to be a problem

after a storm has passed through. It will

cover the solar panels. It will reduce energy

production. Colonists will need to clear the

dust from the panels.

## THE COLONISTS' DIETS

Scientists believe that certain vegetables

can grow in Martian soil. These crops will

provide colonists with nutrients. But many

vegetables are low in calories. They will

not provide colonists with much energy.

Colonists will need to grow other types of

food. Corn, soybeans, and wheat are some

*A plant scientist measures tomato plants that have been exposed to radiation. This will help scientists understand how crops might grow in space.*

examples. These crops can be used to

make a variety of healthy meals.

Colonists will also need protein. It will

be difficult to raise farm animals on Mars.

They take up a lot of space. They need a

lot of food. Scientists are studying ways to grow meat. They think **cells** from farm animals can be used to grow meat. They place the cells in large tanks. They give the cells nutrient solutions. Cells absorb the nutrients and multiply. Scientists are developing ways to make the animal cells have the texture of meat.

Insects are another protein option. They offer lots of calories. Crickets can be ground into a powder. They can be added to other foods. Insects are easy to breed quickly. Raising them also does not take up much space.

On Mars, it will be easier to raise insects for food than large farm animals.

Long-term survival on Mars comes

with many challenges. Humans need to

travel a great distance. They need to be

able to land safely on Mars. Setting up

a colony will require many supplies. The extreme conditions and climate on Mars will cause other problems. Scientists work to solve these problems. With each solution, humankind gets a little closer to building a colony on Mars.

## SLEEP CYCLES ON MARS

A day on Mars is called a sol. A sol is about forty minutes longer than a day on Earth. The time difference can disrupt sleep cycles. This can lead to short-term memory loss. People may be more tired. They may be more likely to make mistakes. But studies suggest that people will eventually adapt to the length of a sol.

Astronauts may one day be able to conduct experiments on Mars alongside rovers and other robotic spacecraft.

# GLOSSARY

**atmosphere**

the layer of gas surrounding a planet or moon

**cells**

the smallest units of an organism

**hemisphere**

half of a planet

**insulation**

material used to shield an object from extreme temperatures

**molecules**

small units of matter made of atoms

**payloads**

things carried by a rocket into space, such as spacecraft
or supplies

**probe**

a spacecraft sent without a crew for the purpose
of exploration

**simulated**

to have artificially produced an object that imitates the original

# SOURCE NOTES

### CHAPTER ONE: HOW WILL HUMANS GET TO MARS?

1. Quoted in "The Challenges of Landing on Mars," *NASA*, December 3, 2003. www.nasa.gov.

2. Steve Swanson, "Are Astronauts Worth Tens of Billions of Dollars in Extra Costs to Go to Mars?" *Conversation*, April 9, 2019. https://theconversation.com.

### CHAPTER TWO: GETTING RESOURCES ON MARS

3. Quoted in Joey Roulette, "Mars Might Be Hiding Most of Its Old Water Underground, Scientists Say," *Verge*, March 16, 2021. www.theverge.com.

### CHAPTER THREE: THE DANGERS OF MARS

4. Quoted in Irene Klotz, "Boiling Blood and Radiation: Five Ways Mars Can Kill," *Space*, May 11, 2017. www.space.com.

5. Quoted in Chelsea Gohd, "There's a Historic Dust Storm on Mars—and It's Nothing Like 'The Martian,'" *Space*, June 14, 2018. www.space.com.

### CHAPTER FOUR: EVERYDAY LIFE ON MARS

6. Quoted in Tom Huddleston Jr., "Elon Musk Says His Mars Base Will Have a 'Fun, Outdoorsy Atmosphere,'" *CNBC*, March 29, 2019. www.cnbc.com.

7. Quoted in Kathryn Mersmann, "The Fact and Fiction of Martian Dust Storms," *NASA*, September 18, 2015. www.nasa.gov.

## BOOKS

Jessica Cohn, *Mars Rovers*. New York: Children's Press, 2022.

Alexandra Lefort, *Discovering Mars*. New York: Applesauce Press, 2022.

Meg Thacher, *The Terrestrial Planets: Mercury, Venus, Earth, and Mars*. San Diego, CA: BrightPoint Press, 2023.

## INTERNET SOURCES

Charles Q. Choi, "Mars: What We Know About the Red Planet," *Space.com*, October 1, 2021. www.space.com.

"NASA Is Recruiting for Yearlong Simulated Mars Mission," *NASA*, August 6, 2021. www.nasa.gov.

Michael Sheetz, "Elon Musk Wants SpaceX to Reach Mars So Humanity Is Not a 'Single-Planet Species,'" *CNBC*, April 23, 2021. www.cnbc.com.

## The Mars Society
**www.marssociety.org**

This organization is dedicated to education and research for the human exploration and settlement of Mars.

## National Aeronautics and Space Administration (NASA)
**https://mars.nasa.gov**

NASA is the US-based agency that is responsible for science and technology related to air and space.

## SpaceX
**www.spacex.com**

This private company designs, manufactures, and launches advanced rockets and spacecraft. It has set a goal of sending humans to Mars.

# INDEX

# IMAGE CREDITS


Cover: © E71lena/iStockphoto

5: © Pavel Chagochkin/Shutterstock Images

7: © Vac 1/Shutterstock Images

9: JPL/USGS/NASA

10: JPL-Caltech/NASA

13: Joel Kowsky/HQ/NASA

16: Kim Shiflett/KSC/NASA

17: JPL-Caltech/NASA

21: © Merlin 74/Shutterstock Images

23: JSC/NASA

25: University of Arizona/Los Alamos National Laboratories/JPL/NASA

27: Norishige Kanai/JSC/NASA

30: JPL/NASA

33: © U3D/Shutterstock Images

35: © Gorodenkoff/Shutterstock Images

37: © Pavel Chagochkin/Shutterstock Images

40: University of Arizona/Texas A&M/JPL-Caltech/NASA

42: JPL-Caltech/MSSS/NASA

45: © E71lena/iStockphoto

47: JPL-Caltech/MSSS/NASA

49: © 3000 AD/Shutterstock Images

53: Cory Huston/KSC/NASA

55: © John Ramkhamhaeng/Shutterstock Images

57: © Gorodenkoff/iStockphoto

# ABOUT THE AUTHOR

Tammy Gagne has written dozens of books for both adults and children, including several titles about space. She lives in northern New England with her husband and son.